STEAL THIS E-BOOK!

An experiment in unsafe texts

by Danny O Snow
with contributions from
Richard Eoin Nash
Dan Poynter
Wade Roush
Glenn Sanders

Steal this e-Book!

is available in paperback or PDF at the following Web location:

www.u-publish.com/stealme

...and is hopefully free to steal as an e-Book from many other sites!

Special Note for Readers in 2008 and following:

This book was originally released in 2002. In the fast-paced worlds of technology and digital publishing, new developments are nearly a daily event. For this reason, *Steal this e-Book!* represents a slice of history, rather than the current "State of the Art" in electronic publishing.

The author has continued to publish related material in the years since this book was released. More recent articles and letters are available at the following Web location:

www.u-publish.com/media.htm

Cataloging Data:

Author: Snow, Danny O.
Contributor: Nash, Richard Eoin
Contributor: Poynter, Dan
Contributor: Roush, Wade
Contributor: Sanders, Glenn
Cover Artist: Kramer, Mary

Title: Steal this e-Book!
1. Publishing 2. Electronic Publishing
3. Computer Technology, Internet

Table of Contents

Acknowledgements

D ANNY O SNOW GRATEFULLY ACKNOWLEDGES the support and encouragement he has received over many years from the late William Alfred, Sam Ardery, Gene Barton, Ted Bayliss, BCS Advertising's Jeanette Brown and Paul Smedberg, Holly Blatman, *BookTech Magazine's* Rebecca Churilla, Kimberly Fabiano, Gretchen Kirby, Donna Loyle and others, Tom Buckner, Mark Butler, the Carey-Davis family, the Childress family, Cynthia (Ling, Ying), Frank DeFord, Don and Sheila, Richard Adrian Dorr; *eBookWeb* founders Wade Roush and Glenn Sanders, Mary Frances England, Sheila Ferguson in spite of it all, Burl Frame, Bethany Gilliland, Bud Gilmore, Michael V.W. Gordon, Judy and Leslie Harrington, John Hartley, Chuck and Robin Harris, John Wm. Houghton, David Jeffers and family, L. Bruce Jones, the Jones cousins, Florrie Binford Kichler, Brandee O'Brien Kingery, Ane Kjølås, Charles King, Larry Larkin, Lo, Chuck Loesche, Rock Lofton, The Clan MacAaron, John Mace, Jack Magestro, Tracy Mayes, Richard Eoin Nash, Cindy Newman, Mark O'Donnell, Kay Olges, *PMA*'s Judith Appelbaum, Jan and Terry Nathan, Dan Poynter, El Scotto and Becky, Robert Burns Shaw, Mike Shartiag and tribe, Dave Shearer, Jane Shore, the Snapp-Childs clan, assorted Snows and Parkhouses, Mark Swisher, Rick Sutton, Greg Temple, Bob Zaltsberg, and most of all, his #1 fans (who also happen to be his parents) Harry David and Jeanne L. Snow. He also thanks thousands of readers of his book titled *U-Publish.com,* co-authored with Dan Poynter, whose feedback has been invaluable in shaping the ideas presented here.

Foreword

— *The French verb "voler" can mean both "to steal" or "to fly."*

G ENTLE READER, Since you're reading this, you are already a participant in this whimsical "experiment in unsafe texts"—and thank you for playing.

This is a retrospective look at electronic books and electronic rights that traces the evolution of e-Books from "ancient times" (1999) to the present. But it may also have an impact on the future. Here's why:

After Dan Poynter and I finished the first edition of our book titled *U-Publish.com* back in 1999, we released it in both electronic and printed form. Ironically, even though the book was about new publishing technologies, the company that distributed the e-Book didn't use encryption at all. I don't believe that this was a deliberate choice; more likely they just didn't care enough to pay for it.

Flattering myself, I wondered if pirated copies would show up on Usenet (they didn't) or if people would buy the paperback that followed in January 2000 (they did).

Dan Poynter wasn't worried. "The book will be out of date within a year since the publishing industry is changing so quickly," he said, and he was right. "I'd consider any pirated copies free advertising for the next edition," he quipped.

In time, we noticed that a substantial number of people who downloaded the e-Book later bought a paperback. That was a pleasant surprise, and it was the beginning of the concept for this book.

After March of 2000, when Stephen King released *Riding the Bullet* exclusively in electronic form, people in the book industry raised quite a fuss about digital rights management (DRM). Unlike my earlier e-Book, King's electronic novella, which was distributed by a real publisher, and used real copy protection, was out for only 48 hours before pirated copies began showing up on the 'Net. And poor Stephen had sold only 400,000 copies at that point.

It amazed me how publishers whined about DRM in the weeks that followed. Jeez, the guy sold almost half a million books with production and shipping costs near zero, and people called it a failure. I longed for that kind of failure.

And I kept thinking about the people who paid for *my* little e-Book with Dan Poynter, and later paid *again* for the paperback, bless 'em all.

According to Glenn Sanders of eBookWeb, in a 1998 article he reported, "Three years ago Rough Guides did the unthinkable; they placed the full text of several of [their] most popular travel guides on the Web," for free. "Ever since Rough Guides placed its content on the site, book sales have increased by at least 20% per year," he added.

More recently, Dan Poynter related yet another example. National Academy Press put an entire series of e-Books online for free ... only to see sales of their tree-Books improve.

Finally, at BookTech 2002 in NYC, other industry experts surmised that e-Books could be good tools to sell tree-Books ... but their comments were more like general impressions than hard facts.

That's when this "experiment in unsafe texts" truly began to take shape. Here's the plan that emerged: a POD publisher suggested compiling a short collection of my writing about electronic publishing, which they would turn into both an e-Book and a paperback. Like the first edition of *U-Publish.com*, the electronic version of *Steal this e-Book!* is going out without copy protection. But this time the e-Book is free, the absence of copy protection is intentional, and we're doing it for a specific reason.

You see, we aren't planning *any* advertising or promotion for the paperback other than the e-book. The free e-Book will be the *only* form of marketing for the tree-Book. Will *anyone* buy the paperback? I don't know—but a year from now we'll have a fairly objective count of how many paperbacks sold as a *direct result* of the e-Book floating around for free.

If you're reading an electronic version and want a printed copy, simply follow a hypertext link to the U-Publish.com Web site and order one; if you don't want a paperback, that's OK too. You're welcome to the e-Book for free.

Personally, I plan to e-mail dozens of copies to friends and acquaintances, and post others at Web sites where readers can download 'em for free. You are invited to do the same, whether you want the paperback or not. So go ahead ... *Steal this e-Book!* As the hacker credo goes, "Information wants to be free." (To which Dick Brass of Microsoft added, "But content providers want to be paid.") Maybe this book will finally prove objectively that both goals can be achieved at the same time.

— DOS
March, 2002

P.S. Did I mention that paperback copies are available at www.U-Publish.com? (Just kidding.)

Preface
by Richard Eoin Nash

RICHARD NASH IS AMONG THE "DIGERATI," the "digital literati" who help today's readers, writers and publishers envision the future of books. A Harvard scholar, author, playwright, and former rights-permissions specialist with Oxford University Press, he also writes about e-Books for several industry trade publications, as well as speaking at publishing events across the nation.

When I first decided to publish a free e-Book to see if it could be a catalyst of sales for a paperback, he was one of the first experts I queried for advice.

Nash is often radical in his views, yet his arguments are compelling. At more than one of his public appearances, I've seen fellow authors and publishers come away with a sneaking suspicion that "we have seen the enemy—and it is us." Happily, the 'sixties-vintage title *Steal this e-Book!* and guerrilla marketing concept held enough appeal to his revolutionary spirit that he agreed to write the preface:

To Buy or Not to Buy, That is the Question
by Richard Eoin Nash

The following headline appeared at www.ditherati.com on 19 February 2002:

EXPROPRIATION TAKES COORDINATION
"Open source means to prove that collaboration works better than authority, or private authorship, for that matter."

— Douglas Rushkoff, predicting that getting people to contribute to his e-book novel for free will help him move units of the print version
Wired News, **19 February 2002**

For the digerati of the world, the fact that e-Books sell print books is so self-evident it's almost beneath contempt.

This book is for everyone else.

This book not only lays out the case for the viability of using wired or wireless technology to deliver content electronically as a branding, marketing, or promotional tool—*it is the case*. This is the metatextual dimension of the book. As with Tristam Shandy and Flann O'Brien's *At Swim-Two-Birds*, there are nested narratives here.

One narrative is the larger debate about what to do with e-Books. Sell them? If so, for how much? Do we let people copy, forward, print, listen, duplicate, select, cut? (The verbs cascade forth.)

A second narrative connects to the evolution of intellectual property in the 30 years since Abbie Hoffman found a new way to announce that property is theft. In particular it concerns the ever-less-material manifestation of intellectual property, especially artistic works. Less and less material is required to store and transmit intellectual property. The container of the idea begins to slowly disappear, leaving only the idea itself, no more palpable than a swarm of electrons ...

... The third is Danny's own narrative, episodes from the last three years of his life as he comes to terms with what the future holds for an industry, and a practice, that he holds so dear. It's the story of the market, of the technology, of the companies, but also the story of the writer, an "Autobiography of an X-Book Man."

The fourth narrative, as I see it, is actually a theoretically infinite number of narratives: the decision each reader (that's you, by the way) will make as to whether to buy the print book. What might your reasons be to buy? To prove a point, pro or con? Because you find it easier to read as hard copy? As a way to "pay" the author, since he'll be paid a little when a print book is sold, but nothing at all for the e-Book? Because a friend might like it and s/he would find it easier to read on paper—although you can, after all, forward an e-copy to as many people as you like? What might your reasons be not to buy? Again to prove a point, this time con?

To buy or not to buy, that is our question to you, and so you too are part of the story this book is telling: "a gift that keeps on giving."

There are no "right" or "wrong" answers to the fundamental question posed by this book: do electronic books compete with printed ones, or complement them?

The electronic version of *Steal this e-Book!* is free; the paperback is not ... and the e-Book is the only form of advertising for its printed counterpart.

It doesn't matter whether ten paperbacks are sold, or 10,000; all of the print sales will come directly from the availability of the free electronic version.

Whether you prefer e-Book, tree-Book or both, you'll be right— and you'll be helping us learn more about what the future holds. With your help, we'll cast this digital bread upon the waters, and see what the tide brings back.

Introduction
by Dan Poynter

D AN POYNTER IS WIDELY RECOGNIZED as one of the world's foremost authorities on independent book publishing and promotion, with more than 80 books in print. In his 'Instant Report' titled *Making the Web Pay* he notes that "... publishers have long been wary of electronic publishing because of a fear of sharing. Having seen all the bootlegging of software, publishers are understandably reluctant to release books as downloadable files that can be copied at the click of a mouse."

Yet the report itself is available to download in PDF format at www.parapub.com—along with hundreds of other reports, documents, and entire books in electronic form. The following excerpts explain why.

Making the Web Pay
by Dan Poynter

There are millions of Internet users and [the] number of people accessing the Web continues to grow every day. It is not lost on publishers that everyone is interested in searching the Web and buying online. In fact, statistics indicate that if you are not using the Internet as part of your business you will no longer be competitive enough to compete in the global digital economy of the 21st Century.

While computer-book publishers are searching for new Internet-Web manuscripts, all publishers are faced with two different challenges: Getting on the Web and making the making money from it.

The Internet is communication channels, and fortunately, publishers have information that can be communicated. We publish what the Web needs: content. Publishers may use the Web to display their catalog of books and to sell those books in both paper editions and in electronic versions online. Customers may be directed to bookstores for the paper version or they may send an order directly to the publisher. Or they can unlock and access an online edition instantly. Now, how does a publisher get people to visit the site and spend money?

Para Publishing has been on the Web since early 1995. The site has continually been expanded with some very clever marketing devices and response mechanisms. This site is an example of what publishers can do on the Web. The site not only shows products and describes services, it sells them. You may wish to log on to www.ParaPub.com to test some of the features as they are described.

For many years, Poynter has been known as an early adopter of new technologies. For example, he was recognized for implementing one of the earliest fax-on-demand systems, and more recently received the Irwin Award for the best electronic promotion campaign by the Book Publicists of Southern California.

Many of Poynter's electronic texts are *not* copy protected, yet his business is profitable. Moreover, he sells both printed and electronic versions successfully. For example, the popular *Self-Publishing Manual* has more than 175,000 copies in print. The availability of the e-Book has *not* eroded sales of the tree-Book.

Poynter's success in marketing electronic texts, starting long before most other publishers, was a big influence in my own experiments with e-publishing, of which this book is the latest example.

Chapter 1: An Experience at OEB
Letter to NPR's "Weekend Edition"

THIS COLLECTION IS CHRONOLOGICAL. When I say "chronological," I mean that it's a series of letters and articles covering events starting in 1999 and continuing to early 2002. They are presented in the order of the events, not the publication dates.

For example, the item below talks about one of the first meetings of the Open e-Book Initiative, early in 1999, though it didn't actually air until the weekend of June 3, 2000.

NPR Commentary on Electronic Publishing

As an early participant in the 'Open e-Book Initiative,' I had the pleasure of meeting with representatives of leading publishing concerns at the headquarters of R.R. Donnelley & Sons in Chicago, early in 1999. The discussion was heady stuff—nothing less than the future of books. At one point, a direct descendant of the venerable R.R. Donnelley himself directed my attention to a panel that slid from the wall, displaying a page from the Gutenberg Bible. The juxtaposition was striking.

As noted ... in your broadcast, these new technologies are still in their infancy, especially in terms of copyright protection. In addition, they are not yet in widespread use by the general public.

Adobe Systems offers special software products named 'Web Buy' and 'PDF Merchant' designed for the secure sale of content from the Internet. Microsoft and Xerox have recently announced the formation of ContentGuard Inc., which promises to allow a document's author, publisher, distributor or seller to secure it against piracy, track its movements, and require users to pay before using it.

However, as noted in your interview, Stephen King's electronic novella *Riding the Bullet* survived less than 48 hours, before pirated copies started to surface on the Internet. According to *The New York Times*, the May 23 announcement about the release of Michael Crichton's thriller *Timeline* and other titles for the Pocket PC was made "even if it is not clear yet how protected the electronic titles are from hackers."

In 1999, the first generation of hardware devices specifically designed for reading electronic books (the Rocket e-Book, SoftBook, GlassBook, etc.) became available to public. At present, however, compared to millions and millions of desktop and laptop computers, the number of dedicated e-Book reading devices in use is extremely limited. The new Pocket PC with Microsoft Reader holds the promise of bringing e-Books more squarely into mainstream markets—but again, it will take time before the number of Pocket PCs even begins to approach the ubiquity of the desktop or laptop computer.

Why, then, are major publishers jumping on the e-Book bandwagon?

The answer is simple: the economic advantages of e-publishing are so compelling that the New York houses can no longer ignore them.

By drastically reducing the physical expenses and economic risks that have traditionally been borne by publishers, electronic distribution will change the entire dynamic of what 'publishing' means in the new millennium.

Eliminating waste and slashing production costs will change the publisher's focus from 'playing it safe' with commercial material, to a new era of innovation and creativity that benefits readers and writers alike.

No one knows exactly what the future holds, but it seems certain that e-publishing is here to stay—and that it will dramatically alter the way writers and publishers reach readers in the 21st century.

The italic text above was broadcast by NPR's "Weekend Edition" over the weekend of Book Expo America 2000. A sound byte is available at the Web location below:

http://www.npr.org/ramfiles/wesat/20000603.wesat.04.ram

Chapter 2: Cancel My Subscription

Harvard Magazine, January 2000

THE NEXT ITEM is a letter that appeared in *Harvard Magazine*. Incidentally, it's published both in print and online, and Harvard warns its alumni that anything they publish in the magazine will be freely available online to the teeming millions. Personally, I like it when more people read things I've written, so this policy is fine by me.

You can't write about electronic publishing without discussing electronic rights at least in passing. The funny thing is that the conversation seems too often to focus on hardware and software issues.

My letter glossed over the issue of copyright protection because it was written in response to an article about hardware. I was confident that a DRM solution would be found, but didn't know what it would be. I still don't.

The "killer ap" for e-Books remains elusive, and I still doubt that the answer lies solely in hardware or software. Instead, my hunch is that the solution will be a combination of hardware, DRM and new pricing and business models that fit the normal buying behavior of consumers.

Sure, an "honor system" may sound naïve, but perhaps not entirely so. Look at the software industry: some shareware developers do make money.

Using a similar marketing model, Adobe distributes the Acrobat reader for free, but users are encouraged to upgrade to the inexpensive pro version.

On the other hand, I hear that there's a brisk trade in black market copies of more expensive software products.

I think that cost is a big factor. When the price is low, consumers will pay for more ease of use, and more features. When the price is high, they are more likely to look the other way at piracy.

Even today in 2002, some publishers persist in charging high prices for e-Books. It's hard to understand why, when the production and shipping costs are so low.

This undermines a primary power of e-publishing: the potential to charge consumers less, pay content creators more—and still make money.

But the power is there, for those who find ways to use it effectively. The right combination of hardware, software and business model will appear in time. The real issue is WHEN, not IF.

Cancel My Subscription

The sheer economics of electronic publishing virtually guarantee that a substantial portion of all publishing will be electronic in the future. By drastically reducing the physical expenses and economic risks traditionally borne by publishers, electronic distribution will change the entire dynamic of what "publishing" means in the new millennium. Eliminating waste and slashing production costs will change the publisher's focus from playing it safe with commercial material, to a new era of innovation and creativity that benefits readers and writers alike.

Jerome Rubin '46 ("The New Gutenberg?" May-June, page 85) is dead on target in his statement that the weak link in the chain of delivering "content" (books, magazines, newspapers, and more) from writers to publishers to readers electronically is the "user interface" (read: computer screen) where the content is read. The publishing industry has made huge and rapid strides in developing software solutions for the delivery of online content, yet the hardware lags behind.

Technologists and publishing-industry watchers now speculate endlessly about which hardware and software will ultimately prevail in the marketplace, how they will work, how they will protect the copyrights of authors and publishers, and a variety of other issues. But it seems certain that e-publishing is here to stay--and that it will dramatically alter the way writers and publishers reach readers in the twenty-first century.

Like Rubin, as much as I would prefer to save Harvard Magazine the cost of printing and mailing each issue to my snail-mail address, it simply isn't comfortable to read the entire magazine while sitting upright before a computer screen. As much as I enjoy your publication, I look forward to cancelling my subscription (to the printed version) as soon as a more satisfactory medium for reading it electronically is available.

This item is still available online in the *Harvard Magazine* archive:

http://www.harvard-magazine.com/archive/00ja/ja00.letters.html#cancel

Chapter 3: Stop the Presses!

Excerpt from *U-Publish.com*
First paperback edition, January 2000

THE FOLLOWING ITEM is a chapter titled "Stop the Presses!" from the first edition of my book titled *U-Publish.com,* co-authored with Dan Poynter, written in 1999.

For context, note the 1999 estimate of the Internet population at "70 to 100 million users," and other quaint little artifacts like the reference to PDF Merchant as a "new" software product from Adobe.

More interesting was the prediction that 100% copy protection might prove impossible. Remember that this was written months before Stephen King's *Riding the Bullet* was released in "secure" PDF format— and cracked within hours.

Mr. Poynter and I emphasized that the best strategy is to *deter* piracy, by making the benefits of fair use (and ease of use) outweigh the savings from stealing a modestly-priced product.

Stop the Presses!

What's important about a book? Does it make a big difference whether the book is printed on white paper or tan paper? Whether the book is 8.5x11" or 5x8" or another size?

While there are a small number of cases where physical appearance is really important, such as picture books for your coffee table or leather-bound classics for your library, usually *it's the words in a book that matter most* to the overwhelming majority of readers.

This is not to say that an attractive book isn't better than an unattractive one, or that an attractive cover and good book design aren't factors in sales. But generally speaking, *people buy books because they want the information contained in them.*

With the rise of the Internet, it is now possible to deliver information anywhere in the world in a matter of moments. The kinds of information now available on the World Wide Web are almost limitless, including not only Web sites *per se,* but also online newspapers, "e-zines" (electronic magazines) and electronic books.

For those unfamiliar with the term, "e-Books" are not physical objects made of paper and ink; instead, they are full length books that can be downloaded from the Internet directly to the computers of an estimated 100 million or more people around the world.

e-Books are also called virtual books, online books, digital books, and a variety of other names. No matter what you call them, they are revolutionizing the entire publishing industry.

Advantages of e-Books:

"Digital books make sense: By eliminating paper and ink, over-the-road shipping, unsold copies and middlemen, Web books sidestep the considerable environmental and economic costs of conventional publishing. Those savings are passed on. Authors typically receive royalty payments of 30 to 50 percent, compared to a conventional industry standard of 5 to 15 percent. Readers come out ahead, too, paying 25 to 50 percent less than softcover prices for most digital books."

-- Cate Terwilliger in the *Denver Post,* 2/99

When a book is published in electronic form, the publisher drastically reduces almost all of the expenses discussed at the beginning of this chapter that create economic risks: printing, binding, packaging, distributing, shipping, warehousing, inventory, percentages paid to middlemen, and returns of damaged or unsold copies.

The last item is the most powerful one. Because e-Books are generated "on demand" (that is: one at a time, as each copy is purchased) there are no wasted copies. At the same time, an *unlimited* number of copies is available to the public. By definition, e-Books are never "out of stock." There is always exactly the right number of copies available: one for every reader, not more, not less.

Electronic books also have powers far beyond those of mortal books. For readers with vision problems, type sizes can easily be increased. Books on subjects that change frequently can be quickly updated, without reprinting. Most e-Books are fully searchable. A library patron will never find that an e-Book is unavailable because someone else has checked it out, nor will there be a late fee for returning it after it is due. A thousand e-Books can be stored in less space than a typical cookbook. Students can copy and paste key passages from most e-Books to book reports without retyping. e-Books can include sound, animation, interactive graphs and charts, and links to online resources.

For example, if you are reading the electronic version of this book on a computer with an active connection to the Internet, you can simply click on the link below to visit the homepage for this book for regular updates:

www.u-publish.com

Technical capabilities aside, the economic advantages of e-Books are why they are turning the publishing industry upside down. Because the cost of bringing books to market is slashed, the publisher's financial risks are virtually (pun intended) eliminated. Because economic risks are nominal, publishers can take a chance on books that might not otherwise reach the reading public. More choices for readers means more books sold. More books sold means lower prices, and lower prices mean more are sold.

Instead of investing $8,000 or more on an initial press run, a publisher can now make an e-Book available worldwide for about one tenth of that amount. If the book sells for $5, the publisher needs to sell only about 200 copies to recover his initial investment in full. Since there is little difference in the publisher's cost to sell 200 e-Books or 200,000, every copy sold thereafter creates a profit.

When economic risks are eliminated, the entire dynamic of publishing a book changes. The focus shifts from "playing it safe" with writers and subjects that have proven commercial potential, to making more choices available to readers, so there is something for everybody. It also allows publishers to take a chance on a greater variety of material, to charge less for books, and to pay writers a larger share of the profits. Everyone wins: the reader, the writer, and the publisher.

Disadvantages of e-Books:

From a strictly technical or economic perspective, e-Books are vastly superior to conventional ones. But they are not without drawbacks.

First of all, not everyone in the world has a computer yet. Although the number of Internet users is huge (estimates range from 70 million to 100 million or more) and growing daily, the fact remains that there are millions of readers who don't own computers, aren't online, or both. Savvy writers and publishers won't ignore these more traditional folks.

Secondly, many people find it uncomfortable to read electronic books on their personal computers. Given the size of most computer screens, an entire page usually won't fit on a computer screen, unless the monitor is very large or the type is extremely small.

A new generation of electronic devices specifically designed for reading e-Books, with names such as the Rocket e-Book, the SoftBook, the GlassBook, and the EveryBook are just now reaching the market in the summer of 1999.

These e-Book readers are smaller than a laptop computer, usually only 2-3 pounds, and their screens are perfect for reading. With these devices, you really *can* comfortably curl up in bed with a good e-Book! As an added bonus, they're backlit so you can read without a light on, too.

It seems likely that lots of them will appear in households in the years ahead. At this time, however, there only several thousand e-Book readers in circulation, compared with millions and millions of personal computers—so for now, most e-Books need to be read on desktop PCs, with the limitations described above.

Of course, it's possible to print an e-Book on your laser printer, but the resulting hard copy loses its search capabilities and many other features that make e-Books special.

Publishers also have security and copyright concerns about electronic books that deserve consideration. After all, they don't want a "pirate" to put a copy of an e-Book on a public Web site, and offer unsuspecting visitors illegal copies without paying for them.

Distributors of online content are making huge strides in the protection of intellectual property. To cite just one example, a new software suite named Web Buy and PDF Merchant from Adobe Systems holds the promise of making it nearly impossible to make unauthorized copies of e-Books, for all but the most determined pirate.

Keep in mind, this book costs less than $10. How much time would you be willing to waste, to steal something with such a reasonable price? We do hope the information proves much more valuable to you than $10—but it probably isn't worth hours and hours of your time, and the risk of prosecution, to steal it.

To skeptics who are still paranoid about pirating of electronic books, we ask this question: suppose an unethical reader buys a copy of any conventional book from any conventional bookstore, then scans it and puts the resulting file anonymously on a public Web site and offers free illegal copies—how can publishers prevent this from happening?

The truth is: they can't. While the publisher can sue if the culprit is caught, our point is that there are no 100% solutions. It's similar to buying a security system for your home: if a burglar really wants to break in badly enough, he probably will. The key is *deter* the crime, making the cost of the theft outweigh the value of what is stolen. In our view, e-Books are already at least as secure as conventional books, and rapidly becoming more so.

Nevertheless, the *perception* can be more important than the *reality*. In the immediate future, the security of electronic books will continue to be controversial.

For the author/publisher who distributes from his own Web site, attracting readers and handling technical issues can be a challenge. Few writers have the computer experience, or inclination, to become full-time webmasters:

"But why not just set up your own Web site and sell your book there? Because, virtual book publishers say, people are a lot more likely to visit a site that has hundreds of books than a site that has only one or two. And because publishers have the resources to promote their site and their books. And because they take care of the hassle. They handle the orders, the credit-card numbers, the downloads. All you have to do is wait for the royalty check."
— Soyia Ellison in the Winston-Salem *Journal,* 9/9/98

One final drawback to electronic books: in most cases, the reader needs a credit card to buy them. Most e-Books are distributed from Web sites, which require the reader to input their card number before they can download the file.

Some readers don't have credit cards. Others are simply reluctant to use them on the Internet.

It seems certain that in time, the general public will be more comfortable using credit cards on the Internet, or that some other form of "cyber-cash" will eventually be used by meaningful numbers of ordinary people.

Meanwhile, the best plan is to combine the benefits of electronic distribution with those of more traditional methods. Happily, such a combination is already available, and gives the writer/publisher an unbeatable one-two punch that knocks the socks off any method available in the past.

Since 2000, we've published two more editions of the book titled *U-Publish.com* to help readers keep pace with rapid changes in technology and the book industry. Many sections of the book are updated online in between editions, at the Web site named for the book.

Chapter 4: Hacking the Bullet

eBookNet, March 2000

e BOOKNET BROKE THE STORY that Stephen King's electronic novella *Riding the Bullet* had been pirated within 48 hours of release. eBookNet later evolved into eBookWeb, now a leading online resource about electronic books and devices for reading them. Founders Glenn Sanders and Wade Roush e-mailed the news to me before the story hit the national wires, with the following item appearing at the U-Publish.com Web site the next day.

Hackers Crack Stephen King's e-Book

Since "U-Publish.com" was released in January '00, new information has become available in some important areas:

In the early chapters of the book, the authors discuss the advantages of electronic books in detail, as well as drawbacks.

Security concerns are among the most important. New technologies, such as Adobe's Web Buy and PDF Merchant software and the upcoming Microsoft Reader, hold the promise of making it possible to distribute electronic books online, while protecting the writer's copyright. If you are planning to publish an e-Book, it is crucial to make sure that adequate copy protection is used.

Even with "industrial strength" encryption, security can present a problem for authors of books with widespread public appeal. As reported on 3/23/2000 by eBookNet, a leading online resource center for electronic books, a major development that illustrates the issue has just occurred. Click on the link below for the full article titled "Cracking the Bullet: Hackers Decrypt PDF Version of Stephen King e-Book" by Glenn Sanders and Wade Roush:

http://web.archive.org/web/20000620153828/www.eBookNet.com/story.jsp?id=1671

Here are some excerpts:

"Pirated PDF versions of Stephen King's *Riding the Bullet* have been circulating on the Internet since March 17. While many ISPs have forced members to remove the decrypted files, they are still available from a Swiss site, providing stark evidence of security weaknesses in PC-based e-Book distribution systems. The episode has irked the companies developing such systems, who complain that export restrictions have kept them from using more powerful encryption techniques...

"The developments could temporarily slow the adoption of Adobe's Portable Document Format (PDF) as a common standard for commercial e-Books. It is still uncertain how crackers disabled built-in encryption mechanisms, which are intended to allow only one person at a time—the purchaser—to display a PDF e-Book on a computer screen. But Simon & Schuster and the commercial distributors of the e-Book are trying hard to limit the damage to Mr. King's legal rights, and e-Book industry insiders are equally anxious to fix the apparent security weaknesses exposed by the decryption...

"Some in the industry fear that the pirating episode could give publishers another reason to hesitate before releasing more of their books on open, general-purpose devices such as PCs and handheld computers, which are considered to be more vulnerable to security attacks than closed, dedicated devices. A 1999 study of e-Book security commissioned by the American Association of Publishers concluded that 'Current general-purpose devices do not provide a trusted base for applications since they were not designed from the beginning with security in mind ... No matter what protection the e-Book system provides the content en route, when it is decrypted for display, it is potentially vulnerable to interception.'"

Chapter 5: MS Reader and PDF Go Head to Head

EBookNet, April 2000

ICROSOFT MADE A MAJOR PUSH to capture the fledgling e-Book market in 2000. By releasing an XML-based alternative to PDF, it could use its market muscle to get Microsoft Reader installed on millions of computers (not unlike the bundling of the Internet Explorer browser with the Windows operating system) and move e-Books more squarely into the mainstream of our culture.

While Adobe had an early advantage, due to the many millions of copies of Acrobat already in use prior to 2000, Microsoft had the power and resources to get its competitive product widely and rapidly installed on the computers of consumers—whether consumers thought that they wanted it or not.

Following as it did shortly on the heels of the Stephen King incident, the MSR roll-out included a lot of discussion about DRM issues. But Wade Roush and I suspected that in the long term, the contest would be decided more on the basis of business models than technology. Our hunch was that the availability of content, and the price, would prove more important than the software itself.

Microsoft Reader and Adobe PDF Go Head to Head
by Danny O Snow and Wade Roush

"May you live in interesting times," reads an ancient Chinese curse. For those who follow electronic books, these are interesting times indeed. Major new products specifically designed for delivery of online content have set the publishing industry abuzz, amid a flurry of controversy over early efforts to bring e-Books more squarely into mainstream markets.

Web Buy and PDF Merchant software from Adobe Systems rolled out early in 2000, promising secure delivery of online content across a wide range of hardware and software platforms. Meanwhile, industry watchers are closely following the introduction of the new PocketPC devices and Microsoft Reader, designed to make electronic content almost universally available to the reading public.

On March 14, Simon & Schuster released Stephen King's electronic-only novella *Riding the Bullet*, and received orders for more than 400,000 copies within 24 hours. As the first electronic-only bestseller, the book marked a watershed in the history of publishing. Yet within 48 hours of its release, pirated copies of King's story began to surface on the Internet, raising new questions about how to prevent e-Book piracy.

In this climate of upheaval, we've compiled the following comparison of new products from Microsoft and Adobe.

Any fair comparison of these products must reflect that Adobe Web Buy and PDF Merchant are already publicly available, while the full version of Microsoft Reader with ClearType has yet to be released. For this reason, the amount of information available about the MS Reader is less detailed. It will be possible to make a more meaningful comparison when Microsoft Reader is available (in "mid-2000," according to Microsoft). EBN readers are encouraged to weigh these factors before drawing conclusions about either product.

In order to present a balanced view of both products, EBN interviewed senior representatives from both companies in April, 2000. Jeff Ramos, director of marketing for e-Books, responded for Microsoft. Mark Heisten, former PR manager of ePaper Solutions, and Rebecca Michals, senior PR manager of ePaper Solutions, responded for Adobe.

Below, we list each company's responses to a series of questions from eBookNet managing editor Wade Roush, and guest columnist Danny O Snow, co-author of a new book about the latest publishing technologies titled *U-Publish.com* ... available in both electronic and printed form. Additional notes have been added in a few places where the writers felt that additional commentary might help put the comments of those interviewed in better context.

What are the technical requirements for your product?

Adobe: There are different requirements for Web Buy & PDF Merchant, depending on the product and features used:

Acrobat Reader with Web Buy and Acrobat with Web Buy, Version 4.05 or higher of either product on the Windows or Macintosh platforms.

For File locking and key distribution:

*   Windows NT®, Intel® i486, Pentium® based, or Pentium Pro based personal computer;
*   Microsoft® Windows NT 4.0 with Service Pack 5, running Microsoft IIS 4.0
*   128 MB of RAM (recommended)
*   40 MB of available hard-disk space
*   CD-ROM drive

For Key distribution only:

*   UNIX ®
*   Sun Solaris 2.6 running Netscape Enterprise Server 3.63 or later

Microsoft: Microsoft Reader will operate on desktop and laptop computers running Windows 95, Windows 98, Windows 2000 and Windows NT, as well as on the next generation of Pocket PC devices powered by Windows, using the CE kernel. It will also be supported in the future on purpose-built book reading devices. There are three primary components of Microsoft Reader:

*   The client software with ClearType, which delivers a paper-like reading experience on-screen
*   Tools for the creation and/or conversion of content to Reader

format

- Distribution software that includes digital rights management support as well as industrial-strength distribution capabilities.

How is copyright protection achieved?

Adobe: PDF Merchant encrypts PDF files and allows the encrypter and seller to control permissions for printing, copying, and annotating documents to fit their business model. When content is sold to a customer, the seller has several easy-to-set options for locking content to a user's CPU ID, user ID, local hard disk, or portable media.

We license our encryption technology from RSA technology, which provides the highest level of encryption available for worldwide use. To further enhance security, we leverage industry-leading certificate authentication from GTE Cybertrust.

Microsoft: We will be talking about our digital rights management technologies in the near future. We recognize the importance of digital rights management to owners of content and have invested heavily to develop a system with which we believe they will be supportive. We are quite confident that our solution in this area will find wide scale adoption by authors, publishers and retailers.

Note from eBookNet: Since this interview was conducted, Microsoft and Xerox jointly announced the formation of ContentGuard Inc., which will market digital rights management system based on the Extensible Markup Language. Microsoft has taken a minority stake in the spinoff company (formerly the Xerox Rights Management Group) and says it will integrate ContentGuard technology into Microsoft Reader as well as many of the company's other tools for authoring, distributing, and viewing content.

Like PDF Merchant, ContentGuard's Extensible rights Markup Language (XrML) will allow a document's author, publisher, distributor, or seller to secure it against piracy, track its movements, and force users to pay before using it. Xerox has published an example explaining how this process might work for an electronic book.

At a press conference announcing ContentGuard's launch, Microsoft CEO Steve Ballmer said "If we do this right, it will have benefits for everybody—rights holders, business, the publishing industry, and consumers." However, he added that "Not all of these technologies will be available the day ContentGuard launches. It will take our engineers a while to get them integrated."

Microsoft has also recently announced relationships with R.R. Donnelley & Sons Co. and with Barnes & Noble (both the .com and brick-and-mortar companies). Both RRD and B&N have substantial interests in protecting the intellectual property of authors and publishers, and have likely been coordinating their plans for distributing Microsoft Reader documents around the introduction of XrML.

How does MS Reader compare with Adobe's Web Buy and PDF Merchant?

Adobe: One of the largest obvious differences ... is that the Adobe digital rights management solution is shipping today and tens of millions of potential customers already use Acrobat or Acrobat Reader and can easily and securely purchase content.

Adobe's solution is also cross platform and cross device—today it operates in the Macintosh and Windows environments on the devices that most of us are already using—i.e. your desktop or note-Book computer.

Adobe PDF is the *de facto* standard for the print/publishing industries and most content today is available in PDF (or PostScript) or easily convertible to PDF.

Therefore, our solution is designed to work well with existing workflows with only minimum incremental effort.

Our strategy is based on partnerships with others in the commerce chain that provide a wide variety of industry standard solutions. [We] believe that the only way to get a Microsoft solution is to work directly with them.

Any other solution is based on technology that is not yet commercially available or tested and this creates a great risk in terms of the reliability of the proposed solution, the time market needed to implement a new solution as well as the risk that customers will not adopt it. Acrobat and PDF have been around since the early '90s and there is no question that they work and are widely used.

Microsoft: PDF is a great solution where you need to ensure that a document will faithfully reproduce when printed to a fixed page size. In the future, though, one of the key consumer benefits of e-Books will be the ability of content to dynamically re-flow across multiple devices—allowing consumers to move their e-Books from their laptop to the Pocket PC and then back again, as just one example. Or from a purpose-built reading device to a PC. This was one of the factors which collectively drove the formation of the Open e-Book Authoring Group and the Open e-Book Publication Specification, which has been adopted by a very broad cross-section of the e-Book community, including Microsoft, Nuvomedia and Softbook. Microsoft Reader has been designed to facilitate that kind of "re-flow" across the broadest possible range of devices. We believe that flexibility, coupled with Microsoft's digital rights management solutions, will be a very compelling customer proposition.

Note from eBookNet: From the two companies' marketing statements, it is difficult to discern major differences in the functions provided PDF Merchant/Web Buy and Microsoft Reader with ContentGuard, except that Adobe's system is specialized for PDF documents while Microsoft has said that

Reader will be compatible with any document packaged using the Open e-Book (OEB) format.

The share of the e-Book market ultimately won by each product may depend less on the technology itself than on the partnerships and licensing agreements each company is forming with organizations in the publishing and bookselling businesses.

When do you anticipate that your products will be in widespread use?

Adobe: Most of the customers that we have publicly announced are either e-tailers or service providers that are integrating PDF Merchant in their ecommerce solutions. We are working closely with several publishers and will be announcing specifics as they become available.

Microsoft: Pocket PCs containing the Reader will become available this spring. A Windows and NT version will be available in mid-summer along with the opening of the barnesandnoble.com e-Bookstore.

Note from eBookNet: Since these interviews were conducted, Microsoft's delivery projection for MS Reader has been revised to "summer." On April 19, Microsoft and three major consumer electronics manufacturers (Compaq, Hewlett-Packard, and Casio) announced the availability of new palm-size PCs running Microsoft's PocketPC operating system. A version of Microsoft Reader comes preinstalled on these devices, but eBookNet has been unable to determine whether this version includes copyright protection functions. Fewer than three dozen book titles are currently available for Reader, and all are public-domain works that do not require encryption for copyright protection.

What are the differences between Microsoft's ClearType and Adobe's CoolType?

Adobe: Here is another one of those areas where it is hard to comment because Microsoft has not begun shipping ClearType and our information is limited. The general approach behind the solutions is fairly similar in that they both take advantage of sub-pixel addressing on color LCD screens. My understanding is that Microsoft is focusing primarily on the Windows CE platform and that their technology only works within the Microsoft Reader, and with a limited set (6) of TrueType fonts. In most cases, the author of the document will not be able to use the typeface s/he originally intended, and readers of content will similarly be restricted by not being able to take advantage of the tens of thousands of typefaces available.

Adobe CoolType is platform and device independent and will work with all of Adobe's applications regardless of font type (TrueType, OpenType, Type 1, Type 3, etc., etc.). Authors can retain control over the look and feel of the documents and not be restricted in their choice of operating system. Adobe has 18 years of rendering type to screen and print and the algorithms we have implemented provide the highest quality reading experience commercially possible today.

Given that, it is important to note that an improved reading experience, while very important, is only one key factor. Adobe believes that availability of content from a variety of sources that can be easily read on a variety of devices is much more critical to providing the best consumer experience possible.

Microsoft: [We] do not have technical information on CoolType, but understand it to be remarkably similar to ClearType in principle.

Note from eBookNet: Neither Microsoft nor Adobe invented subpixel rendering, which was first applied to computer monitors by Apple II programmers in the late 1970s. (See Gibson Research Corporation's excellent Subpixel Rendering Web site.) The basic technology is in the public domain. eBookNet's belief is that ClearType and CoolType are essentially equivalent technologies that deliver more or less equal improvements in readability on color LCD screens.

What kind of e-Books can readers expect to find available for these systems? Will any of them be free?

Adobe's PDF is currently the most widely used format for distribution of e-Books worldwide. As mentioned above, Adobe stated that it is "working closely with several publishers and will be announcing specifics as they become available."

Microsoft is currently distributing 29 public-domain titles for Microsoft Reader on a CD-ROM accompanying the new PocketPC devices. Microsoft also stated that a wide variety of content will be available from Barnesandnoble.com and other providers.

EBN's projection is that a large amount of content for the MS Reader will become available through third-party content conversion service providers. This model can be contrasted with a system in which content creators (primarily writers and publishers) release their work directly to the public. The availability of free content is likely to depend the provider and its business model.

Is your system suited for extended reading on a handheld device?

Adobe: One of the nice things about PDF is that you don't need a dedicated device to read PDF files - any personal or desktop computer running the Macintosh, UNIX, or Windows operating system will do! This is extremely important to most consumers as it means that they do not need to buy (or carry) an additional device to participate in the e-Book revolution.

Some of the dedicated devices that are being built are quite nice and I am sure that some consumers will purchase them. Several alternative device manufacturers are implementing PDF solutions such as Everybook, and last month Adobe announced plans to bring PDF to the Windows CE platform, and in conjunction with Palm Computing to the Palm platform. In January, we made a PDF Viewer available for Java. Adobe also publicly announced and showed our plans to enable reflowing of PDF documents to better support different size screens.

There will be additional device manufacturers announcing their support going forward.

Microsoft: The cornerstone of our strategy for the Pocket PC is to give users the power to read whatever they want, wherever they want. If the market wants a device purposed exclusively for reading e-Books, that's what we'll give them; if they want a device that offers a broad range of other applications and capabilities, we can offer that, too. The real power of Microsoft Reader with ClearType is that content providers will immediately be able to address the 150 million PCs that are currently in use worldwide via the MS Reader. That's a huge increase over the current market for dedicated reading devices, and we feel it represents a significant growth element in the emerging e-Book market.

Related Reading

Adobe PDF:
http://www.adobe.com/products/acrobat/main.html

Open e-Book Initiative:
http://www.openebook.org/

Pocket PC Home Page:
http://www.microsoft.com/pocketpc/

Microsoft Reader:
http://www.microsoft.com/reader/

Subpixel Rendering:
http://www.grc.com/cleartype.htm

Extensible rights Markup Language (XrML)
http://www.xrml.org

Chapter 6: Turning Content into Gold

BookTech Magazine, September-October 2000

THE POCKET PC rolled out in tandem with Microsoft Reader. Now Microsoft had both an alternative software product, and a hand-held device to run it.

Unlike earlier efforts to popularize e-Book reading devices, the release of the Pocket PC offered consumers features such as word processing, spreadsheets and e-mail in a hand-held unit, in addition to e-Book reading. Like Microsoft Reader, the Pocket PC also had the market strength of major manufacturers behind it. Finally, it was designed to address some of the weaknesses of previous e-Book readers, adding features such as a color screen and improved text display.

Perhaps most importantly, the release of the Pocket PC triggered a flurry of new entries into the e-Book market by media giants like Random House, Simon & Schuster and Time Warner, later to merge with AOL. Suddenly it seemed possible that large numbers of good books could become available in electronic form, as big media companies followed Microsoft into the market for electronic books.

Yet the Pocket PC was expensive, and sold only modestly. Likewise, major publishers asked high prices for their e-Books, sometimes higher than the printed versions!

When consumers didn't rush to buy millions of Pocket PCs, and didn't download scads of expensive books, the momentum was lost.

Here again, we see the power of consumer choice over technology. Industry giants misjudged the factors that drive consumers, and failed to harness the fundamental strengths of electronic publishing. As we'll discuss later, this led to the collapse of iPublish.com and other e-publishers just a short time later.

The item below is a full draft of an article written for *BookTech Magazine,* rather than the heavily abridged version that appeared in print.

Turning Content into Gold
Special to BookTech Magazine by Danny O Snow

In ancient times, alchemists sought in vain for the mythical "Philosopher's Stone," fabled to transmute base metals into precious ones. The lure of turning lead to gold was irresistible, but the Philosopher's Stone proved elusive, and the alchemists faded away after centuries of fruitless searching.

In recent times, publishers have been equally tantalized by the potentials of electronic publishing: a way to make books available worldwide without printing costs, without warehousing and inventory, without shipping, without returns, without waste. The lure of these possibilities is irresistible to publishers, yet to date, the right combination of hardware, software and marketing to make e-publishing viable has proven as elusive as the Philosopher's Stone.

Enter the Pocket PC with Microsoft Reader, now publicly available: some experts are convinced that the eqivalent of the Philosopher's Stone is now within the publisher's grasp, while others believe that viable e-publishing remains a tantalizing myth. Either way, the release of these new products, and a flurry of important new business alliances related to them, represent an important, possibly historic, development in the history of publishing. This article will explore the strengths and weaknesses of the Pocket PC with Microsoft Reader, as well as a few examples of how publishers are responding to its release.

Hardware & Software:

The term "Pocket PC" applies to a handheld computer with a variety of uses, including reading electronic books, as well as word processing, e-mail, web browsing, audio files, etc. There are currently several such devices on the market, including Casio's Cassiopeia; Compaq's iPAQ; Hewlett-Packard's Jornada, and Symbol's PTT 2700 and run under Microsoft's "Windows-powered Pocket PC" operating system, a mini version of Windows CE. Although it is intended as a multi-purpose device, the Pocket PC is of special interest to publishers because of its potential as a tool for reading electronic books with better performance than earlier products, as explained below.

The Microsoft Reader is software that comes pre-installed on a Pocket PC. It offers a variety of special functions such as highlighting, bookmarks, notes and drawings, search, built-in dictionary, large print, audio books, and others. Microsoft Reader uses new "digital rights management" (DRM) technology from ContentGuard, a feature of great interest to publishers because it promises the possibility of secure delivery of books (and other content) to consumers over the Internet. Microsoft is expected to integrate MS Reader with other major software products soon, allowing electronic books to "re-flow" across a wide range of screen shapes and sizes.

ClearType is special software for LCD screens that provides significantly better readability of text than those used on earlier handheld devices.

Specifications:

- Size: about 3x5"
- Weight: about 6 to 9 ounces
- Memory: 16 to 32 MB
- Processor: 32 bit, 131 to 206 MHz
- Screen: 320x240 16-bit active matrix display with 4,096 to 65,000 colors

While the specifications above show that the Pocket PC packs more powerful hardware than dedicated e-Book readers do, current prices run as high as $500 to $600. Retailers justify the higher cost by citing additional features such as Pocket Word, Pocket Excel, Pocket Outlook, Pocket Internet Explorer, and more. However, at this writing, it remains unclear whether the performance of these other features will meet the demands of consumers. For example, while the Pocket PC boasts a color screen and ClearType software to improve the appearance of text, the display is still only one quarter of the area of the smallest desktop units. At Book Expo America in June, Jeff Ramos, Microsoft's director of marketing for e-Books, drew laughter with his forthright quip that the current generation of Pocket PC devices can be compared to the 286 class of desktop computers.

Earlier e-Book Efforts:

This summer's release of the Pocket PC with Microsoft Reader is of special interest to publishers because it tries to address key problems encountered in earlier efforts to bring electronic books more squarely into mainstream markets. To cite just a few examples:

Some readers found earlier devices less than ideal for pleasure reading, due to their small and colorless LCD screens. The Pocket PC offers a color screen. While still small, it uses ClearType software to enhance the readability of text. Early response from users suggests that ClearType provides a meaningful improvement in the appearance of type, in spite of the screen's modest size, especially when sharp black text is displayed on a white background.

Previously, consumers questioned the value of "dedicated" (single-purpose) devices designed solely for reading e-Books. The Pocket PC offers word processing, e-mail and other functions, though critics claim that some features have been stripped down to fit this small, hand-held device. Although this article focuses on its application to electronic publishing, it is important to recognize that the Pocket PC is marketed as a multi-purpose device, rather than a dedicated book reading product.

Many owners found the delivery mechanisms used to load information on previous e-Book reading devices cumbersome. The Pocket PC is designed to streamline the process of moving information between a consumer's desktop computer and his or her own handheld unit, although critics argue that this may make the information more vulnerable to piracy.

For publishers, a major weakness in earlier attempts to bring e-Books to a broad segment of the public was the failure of copyright protection. Microsoft Reader software uses ContentGuard, a new system of copyright protection that promises to allow a document's author, publisher, distributor or seller to secure it against piracy, track its movements, and (if applicable) force users to pay before using it. While the effectiveness of ContentGuard is still not certain, the ability to sell electronic books without piracy will be extremely attractive to publishers—if it works.

Finally, the total number of previous e-Book reading devices purchased by the public was disappointing to publishers. Between the April 19 unveiling of the Pocket PC and June 7, roughly 10,000 units were sold. While this is still an almost insignificant number to publishers, it seems likely that with the market muscle of Microsoft and other industry giants (see following for details) behind it, the Pocket PC will reach a broader segment of the reading public than its predecessors. Perhaps more importantly, industry experts predict that Microsoft Reader software will soon be integrated into the Windows operating system and/or Microsoft's other leading software products such as Word and Publisher, opening the door for wider use of electronic books by consumers, whether or not they own a Pocket PC.

In combination, the improvements attempted above could signal the arrival of the e-Book as a viable medium for publishers, if the hardware and software deliver the advantages promised by their designers. While it is too early for industry judges to hand down a final verdict on the overall effectiveness of the Pocket PC and Microsoft Reader, its release has sparked a sea change in the behavior of major New York publishing houses, as discussed below.

Major Publishers Jump on the Bandwagon:

For publishers, new corporate alliances involving Microsoft and publishers are probably more significant that the products themselves.

In previous years, major publishers seemed reluctant to embrace electronic books, in spite of their obvious potential to revolutionize the industry. Their reasons, both public and private, varied from copyright concerns to quality control to the understandable fear of losing market share to independent publishers and self-publishers that could result from widespread delivery of books over the Internet.

With Microsoft's entry to the e-Book market, it appears that the dam has started to break. On May 23, new partnerships between Microsoft, Simon & Schuster and Random House were announced with a media fanfare in NYC, including the electronic release of Michael Crichton's thriller *Timeline*, and other e-Books for the Pocket PC with Microsoft Reader.

A major factor in this watershed was almost certainly the prior announcement that Microsoft and Xerox had jointly created a spin-off company named ContentGuard to provide copyright protection for electronic books and other online content. ContentGuard is designed be the primary security feature of MS Reader.

Following the controversy in March over the pirating of Stephen King's e-Book, *Riding the Bullet*, publishers welcomed the ContentGuard announcement, hoping for a "silver bullet" to kill the specter of hacking that loomed over the future of e-Books at the time.

In spite of a warm reception from major publishers, new DRM technologies are still unproven. Ironically, according to *The New York Times*, the May 23 announcements about new e-Book releases from Simon & Schuster and Random House were made "even if it is not clear yet how protected the electronic titles are from hackers." As this article goes to press, the long-term effectiveness of ContentGuard as a deterrent to piracy is still unknown.

Concurrent with the alliances between Microsoft, Simon & Schuster and Random House made public on May 23, Time Warner announced that it will launch a major new electronic publishing venture named "iPublish.com at Time Warner Books" in the first quarter of 2001. iPublish will include a suite of "channels" named iRead, iWrite and iLearn, providing a broad selection of online content, ranging from electronic versions of major best sellers, to works by aspiring writers, plus a number of resources for writers and publishers.

According the Gregory Voynow, General Manager of iPublish.com, iPublish will provide an "online community" for both readers and writers. The iWrite channel will offer an open door policy for aspiring writers, who may submit excerpts of unpublished works for peer review at no cost. The best received iWrite titles will gain further attention from professional editors, and may ultimately advance to commercial electronic release from the iRead channel, or possibly even print publication through one of Time Warner's several traditional imprints, such as Little Brown & Company, Warner Books, or Aspect, to name just a few.

For readers, iRead will feature electronic versions of professionally published books, ranging from major best sellers to re-releases of midlist titles for which traditional reprinting is not cost effective, yet still hold appeal. Nearly 100 frontlist titles will be available when iPublish launches, with hundreds more to be added during 2001, as iPublish outsources selected titles to conversion partners for deployment as electronic books. With combined holdings of 4,000 to 5,000 titles from Time Warner's traditional print publishing subsidiaries, iPublish's potential catalog could ultimately rival or exceed almost any other source of professional quality electronic books.

iRead will offer readers a choice of e-Book formats, including MS Reader, PDF and other formats for handheld devices such as the Rocket e-Book reader and Palm Pilot, which can be delivered to the public natively or through third party content providers.

iRead will also invite readers to participate in an online community that explores their needs and wants, which will influence iPublish's future development, and improve its ability to meet future market demands more effectively. "We expect to be surprised," says Voynow, whose comments to BookTech clearly reflect a long-term perspective on how books will be published in the new millennium.

iPublish's future could also be strengthened by an impending corporate relationship between its parent, communications giant Time Warner, and America Online, the world's largest internet service provider ...

... While the sheer economics of new technologies are virtually certain to drive publishers more and more in the direction of e-publishing in years to come, these technologies are still in their infancy. In the long term, it seems likely that the Pocket PC with Microsoft Reader will be remembered more for opening the door to serious alliances between publishers and computer manufacturers, than for the hardware and software *per se*.

However, the opening of this door is in itself a meaningful development. Unlike the earlier efforts of young high-tech upstarts to popularize electronic books, publishing giants like Simon & Schuster, Random House and Time Warner have the resources to attract and meet large-scale consumer demands. They also control large catalogs of good books for consumers to read, compared to the relative scarcity of top-quality material available from early e-Book proponents. As is often the case with new technologies, it appears that a once-revolutionary concept will now start to be co-opted by powerful industry players, as the technologies begin to mature.

Whether the release of the PocketPC with Microsoft Reader represents a giant step forward on the map of publishing's electronic future, or only a modest stretch of new ground, events in the summer of 2000 may well be remembered for bringing the end of the road closer to view. The Philosopher's Stone remains elusive, but is drawing closer to the publisher's grasp.

Related Reading

Pocket PC Home Page:
http://www.microsoft.com/pocketpc/

Microsoft Reader:
http://www.microsoft.com/reader/

ClearType:
http://grc.com/cleartype.htm
http://www.microsoft.com/reader/ppc/product/cleartype.htm

ContentGuard:
http://www.contentguard.com

iPublish.com at Time Warner Books
http://www.ipublish.com

Casio's Cassiopeia:
http://www.casio.com

Compaq's iPaq:
http://www1.compaq.com

HP's Jornada:
http://www.hp.com/jornada/products/540/prod_spec.html

Microsoft, ClearType and Windows are either registered trademarks or trademarks of Microsoft Corporation in the United States and/or other countries. CASIO and CASSIOPEIA are registered trademarks of CASIO Computer Co., Ltd.

Chapter 7: e-Book Formats Spar and Parry

Internet Publishing Magazine
February 2001

THE BATTLE FOR SUPREMACY in the e-Book world between Microsoft and Adobe continued throughout 2000 and into 2001. The article below discussed differences in the strategies used by each side for presentation of content to readers in electronic form.

Meanwhile, both camps beefed up encryption in an effort to satisfy publishers that their intellectual property was reasonably secure. After the Stephen King incident, pirated e-Books faded from the headlines. But tighter security required more complex systems, often forcing consumers to install new software—which didn't help retail sales of e-Books in the slightest.

e-Book Formats Spar and Parry

Some technology battles are classics, such as when VHS KO'ed Beta to emerge as undisputed champion of the VCR world. Now, a new rivalry has emerged in the fledgling, but very hot, e-Book arena.

The two leading e-Book formats are Adobe's Portable Document Format (PDF) and Open e-Book (OEB) formats, such as LIT files for the Pocket PC with Microsoft Reader, reported Danny O Snow of Unlimited Publishing LLC, BookTech advisory board member, and a noted authority on new publishing technologies.

The tale of the tape reveals each format's strengths and weaknesses. "PDF's great strength is the faithful reproduction of the printed page, as the book designer created it," Snow said. "PDF files are basically digital pictures of printed pages that appear virtually identical —provided that they are viewed on a screen of roughly the same proportions as the originals, which is not always the case."

On the other hand, OEB formats tend to roll with the punches. "The power of OEB formats is precisely the opposite," he continued. "Their ability to re-flow text across screens of varying shapes and sizes makes OEB files compatible with a wide array of hardware devices--including some that haven't been invented yet."

In short, PDF's aesthetic consistency may please content creators, while bean counters may value more highly OEB's open nature--compatibility with more distribution vehicles.

"Writers, editors, designers, artists and production people tend to favor PDF, because it affords greater control of presentation to the reader," Snow added.

"On the other hand, many marketing and business people feel that the potential to sell more e-content to more consumers is more important than details like type kerning, widows and orphans, or a river of white space running through text."

Preparing for the next round, both contenders are working to improve their conditioning. Snow related that Adobe is working steadily to improve PDF's ability to re-flow, while predicting that future OEB-compliant formats are sure to look better and better in the years ahead.

Related Reading:

Adobe PDF:
www.adobe.com/products/acrobat/main.html

Open e-Book Initiative:
www.openebook.org

Pocket PC Home Page:
www.microsoft.com/pocketpc

Microsoft Reader:
www.microsoft.com/reader

Subpixel Rendering:
www.grc.com/cleartype.htm

Extensible rights Markup Language (XrML):
www.xrml.org

eBookWeb:
www.eBookWeb.org

Chapter 8: Tree-Rights not e-Rights

Published at the U-Publish.com Web site
July 2001

NEW METHODS FOR DELIVERING CONTENT to consumers in electronic form continued to evolve at the dawn of the new millennium. In this context, the word "content" includes books, music, newspapers and magazines, and even early online movies.

As ever, the worldwide scope of the Internet fostered new and creative business models. New technologies made it possible to re-package, re-sell, and re-deliver all kinds of content, in ways that were impossible just a few years ago.

The downside was (and still is) that innovative marketing and delivery methods raised questions about ownership of intellectual property. Copyright laws enacted in the "analog" (pre-digital) era were not designed to deal with the kind of flexibility and seamless transmission of content that became possible with the rise of the Internet. In 2001, important disputes about electronic rights reached the courts. The Napster case, *Tasini v. New York Times* and *Random House v. Rosetta Books* were among the most prominent.

Specifics about these cases are widely available elsewhere (see the following links for just a few examples) but in general the courts held that electronic versions of various forms of content must be treated separately from their antecedent pre-digital forms. In other words, before turning a printed book or article into an electronic one, you need permission from the person who wrote it.

The summary above is of course over-simplified; arguments about ownership of intellectual property are complex. But the central questions remain: *Is it possible to make money in an environment where most things are free, and if money IS made, who gets it?*

In my view, the eventual solutions won't come only from the courts, any more than they will come solely from software and hardware developers. Instead, I think the answers will evolve from new pricing and business models that are more compatible with the normal behavior of consumers.

Supreme Court Rules in Favor of Writers
Tasini v. New York Times

The Supreme Court has ruled that print publishers such as newspapers and magazines may not re-use material online without paying the writer in cases where they had previously obtained only print rights. The Court's ruling establishes that online and electronic rights are separate from print rights.

According to the National Writers Union: "By a 7-2 majority (Stevens and Breyer dissenting), the Court upheld a September 1999 unanimous ruling by the U.S. Court of Appeals, 2nd Circuit, which found that *The New York Times* and other publishers had committed copyright infringement when they resold freelance newspaper and magazine articles, via electronic databases such as LexisNexis, without asking permission or making additional payments to the original authors."

Chapter 9: e-Rights Update

PMA National Newsletter
October 2001

WHEN THE COURTS SHUT DOWN NAPSTER, the big media companies—including major book publishers—celebrated. Finally, they thought, at last the door was closing on free downloading of copyrighted material from the Internet.

The party was premature. Within weeks, millions of college students resumed the practice of online music "sharing" at other sites, often using new software that didn't require a centralized registry like Napster.

In the book world, the *Tasini v. New York Times* and *Random House v. Rosetta Books* cases dealt big publishers additional blows, establishing that they had to pay writers for electronic rights. (Random House's suit is still in progress; I predict it will reach the Supreme Court.) Now publishers felt squeezed from both sides.

In the following article for the national newsletter of the Publishers' Marketing Association, I argued once again that the solution lies in new pricing and business models, rather than the courts or technology: "The fundamental economics of electronic publishing make it possible for publishers to charge consumers less, and pay writers more."

After a publisher has prepared a book for traditional print distribution, I doubt that the per-unit cost of distributing electronic copies is much more than a nickel a copy. Why not make them affordable?

And why not try new business models? Early Internet services such as Prodigy and AOL charged by the hour. Consumers balked, and today's "all you can eat" (unlimited use) model evolved. Wouldn't the same kind of model work for an online library of quality material?

Likewise, years ago, many telephone companies tried to establish "measured service" for local calls (read: pay by the call) as the norm. NYC residents currently pay for each local call, but it's cheap, maybe five or ten cents a call, and the monthly bill is about the same as unlimited local calling plans elsewhere.

Either plan might work for e-Books: the single pay-per-read (if inexpensive) or the "all you can eat" subscription model.

"How can we make money selling any book for a dime?" asked one colleague who works for a big New York publishing house. "It depends on how many books you sell," I replied. "There's no cash difference between a hundred thousand dollars and a million dimes. I guarantee you'll sell more books for a dime than for a dollar—and what's wrong with more readers?"

My colleague then expressed a common fear among publishers: people will stop buying tree-Books if they can get e-Books for a fraction of the cost.

But most of what I've learned suggests that the opposite is true. First, even when both are available, most printed books outsell their electronic counterparts by at least 10-to-1. Next, there are reports such as those mentioned earlier about Rough Guides and National Academy Press, which suggest that e-Books actually *increase* the sales of tree-Books. While these earlier reports are anecdotal, we intend to pay close attention to sales of *Steal this e-Book!* in printed form. This should shed clearer light on how much e-Books catalyse sales of printed ones.

Those of us who make our livings with words may not like the public's penchant for passing along our words without paying us directly. Unfortunately, we don't get to make the rules, nor do the courts or the computer companies. The public is why we all exist, and they're the ones who will ultimately decide how the game is played.

The lessons of history are clear: cassette tape recorders didn't kill the music business, nor did Xerox copy machines kill the book business. Other examples are discussed below. The point is that we authors and publishers (not the public) must be the ones to change and adapt to the needs and wants of readers ... not the other way around.

e-Rights Update
Special to the Publishers Marketing Ass'n by Danny O Snow

Throughout 2001, conflicts pitting established media giants against high-tech upstarts have focused the attention of publishers on new technologies and e-rights. This report reviews highlights of recent developments and concludes that the dangers of new technologies that bypass publishers of books may be overstated.

Napster Falls...
But Will "Bookster" Rise from the Ashes?

Although it centers on music rather than books, the Napster case is important to authors and publishers. That's because the kinds of technologies used for trading music online are already beginning to be applied to electronic books.

Napster has, of course, generated reams of general coverage in major print media. As it relates to book publishers, one of the most alarming aspects was reported by Columbia University law professor Eben Moglen in *The Nation* on March 12, 2001: "The shuttering of Napster will not achieve the music industry's goals because the technology of music-sharing no longer requires the centralized registry of music... that Napster provided. Freely available software called OpenNap allows any computer in the world to perform the task of facilitating sharing; it is already widely used."

For publishers, this raises the specter of a Napster-like online sharing site for electronic books that might emerge in the future, using OpenNap-style technologies. Some e-Books can already be found on peer-to-peer networks such as Gnutella. Moreover, some of today's leading solutions for e-Book copyright protection are intentionally designed to allow purchasers of e-Books to lend single copies legally.

According to Dian Killian of the National Writers Union, "It all comes down to fair use. Before the Internet, no one cared if you loaned a music recording or book to your family or a small circle of friends. With Napster and e-Books, it's now technically possible to 'loan' a recording or a book to thousands and even millions of strangers."

Electronic delivery of content in various forms is driving the development of new and creative ways to wring additional revenues from intellectual property every day. The downside is that innovative marketing and delivery methods are also raising new questions about ownership of rights, and compensation for both the author and the publisher.

Tree-Rights not e-Rights

As discussed by Jonathan Kirsch in September's PMA Newsletter, recent court rulings in lawsuits (*Random House v. Rosetta Books* and *Tasini v. New York Times*) establish that online and electronic rights are separate from print rights.

These decisions need not pose serious problems for most publishers. They simply reinforce the need to schedule specific rights in publishing contracts, and to pay for them as appropriate. Many publishers already include electronic rights in their contracts, or simply secure "all rights" when they acquire new titles. Electronic rights for earlier titles may be negotiated on a case-by-case basis if necessary.

How the Book World Has It Better

"Liberation Musicology"—the article about Napster in *The Nation*—concludes with what sounds like an ominous portent for publishers: "What is most important about this phenomenon is that it applies to everything that can be distributed as a stream of digital bits by the simple human mechanism of passing it along. The result will be more music, poetry, photography, and journalism available to a far wider audience. Artists will see a whole new world of readers, listeners, and viewers."

But let's not forget the lessons learned in the music and film industries during the 20th century. When cassette recorders appeared, cynics claimed that the record labels were dead. Likewise, some said the VCR spelled the doom of [movie theatres]. Neither prediction proved true, and in fact, the home video market is now a major source of revenue for filmmakers. The Internet raises the ante, but the publishing game remains the same.

And let's also not forget that publishers can have better relationships with readers and writers than the recording industry has with its artists and customers. Record companies are in danger for a couple reasons. For one, their artists feel alienated (given the chance to jump ship, many did so without hesitation). Also, their consumers have resented the huge markup on CDs (often many times the margin on a printed book), and some listeners rationalized online music "sharing" as a chance to recoup their losses.

The fundamental economics of electronic publishing make it possible for publishers to charge consumers less, and pay writers more. Publishers who adopt this philosophy may reduce the risk of piracy by readers, and increase the loyalty of writers—hopefully without costly litigation, before or after the fact.

Keeping Readers Honest

As "brick and mortar" bookstores know, a small number of consumers have always stolen books. Losses from theft are a standard factor in calculating bookstores' operating expenses. Books with high prices seem more likely targets of shoplifters—online or offline.

When I chatted about electronic rights recently with three bright young graduate students, they gave me a familiar opinion—paying for online content is anathema to the spirit of the Internet. Pressed on the issue of fairly compensating writers and musicians for their work, though, the students allowed that a modest payment for use of online books and music could become the norm.

My experience [at] Unlimited Publishing LLC reveals a related pattern. We publish books primarily in printed form, but recently began releasing e-Books [of selected titles.] We typically price the electronic editions below $5, while paperback prices are $11.99 to $22.99. In the planning stages, we learned that a substantial number of consumers…who download an e-Book later purchased a printed copy. As a result, we now view e-

Books as good tools to sell tree-Books… not much different than free review copies given to journalists and VIPs.

Keeping Writers Loyal

Many conventional publishers admit privately that their relationships with writers are adversarial to some degree. Perhaps this is because many writers perceive their publishers as greedy, while few recognize that part of the income from every book sold must offset losses from books that *don't* sell.

Consumer pilfering of books (whether printed or electronic) pales in comparison to the impact of returns of unsold copies for most publishers. Theft may account for a small percentage of losses—but return rates from bookstores of 25% to 35% or even higher are not uncommon. Interestingly, new technologies—such as Print on Demand and e-Books—can dramatically reduce returns numbers, practically eliminating the publisher's most painful problem. As a result, high-tech publishers can pay writers more, whether the book is made of paper and ink, or bits and bytes. For instance, we pay at least 50% of net from retail sales of printed books to clients, and [the e-Book distributor] pays 76% of revenues from e-Book sales. Policies like these foster better relationships with writers.

What's Ahead?

While copyright protection for e-Books is still a concern, the potential to deliver books to consumers without waste may strike many publishers as nearly irresistible—whether the book is printed on demand or downloaded from the Internet. The price publishers must pay for the promise of these technologies is a willingness to change. Whether this means developing new production workflows, embracing new pricing and royalty models, or both, the benefits can outweigh the costs for the publisher of the 21st century.

This article is expanded from items published previously in print by BookTech Magazine and online by eBookWeb, with excerpts used by permission.

Related Reading

"Liberation Musicology"
Eben Moglen on Napster
http://www.thenation.com/doc.mhtml?i=20010312&s=moglen

National Writers Union
Publication Rights Clearinghouse:
http://www.nwu.org/prc/prchome.htm

Publishers Weekly on Random House v. Rosetta Books
http://www.publishersweekly.com/index_articles/20010305_94700.asp

Random House
http://www.randomhouse.com

Rosetta Books
http://www.rosettabooks.com

U.S. Supreme Court
on *Tasini v. New York Times*
http://www.supremecourtus.gov/opinions/

The PMA newsletter in which the entire article appeared is also available online at:

http://www.pma-online.org/scripts/shownews.cfm?id=556

Chapter 10: iPublish.com and MightyWords Fold

Published at the U-Publish.com Web site

B Y THE END OF 2001, major publishers like Random House began pulling the plug on electronic books. The events of 9/11 and a general downturn in the U.S. economy were big factors—but not the only ones.

Sales of e-Books, and devices for reading them, remained modest; estimates vary, but many industry observers believe the number of dedicated e-Book reading units "on the street" had grown to only 100,000 or so. For publishers who want to sell millions of books, this was not a big enough market to justify additional investments during a recession.

Poor marketing was also a factor. Even today at some online retailers (most notably Amazon.com) e-Books are still segregated from tree-Books for some unfathomable reason. Imagine if paperbacks were listed separately from hardbacks, or if consumers who wanted audio tapes were forced to search an entirely different catalog. Whether this is just bad planning, or outright prejudice isn't clear ... but it certainly didn't help the market for electronic books.

Legal issues like those discussed earlier posed additional disincentives to publishers who had formerly planned to re-release large numbers of good books in electronic form.

The modest number of e-Book reading units in use, combined with a reduction in the number of quality books available to read on them, created a "Catch-22" for both readers and publishers. On one hand, the market wasn't big enough to give publishers a strong incentive to release thousands of titles in electronic formats; on the other hand, the relative scarcity of quality titles didn't give the public an incentive to buy e-Book reading devices.

As a result, the following reports appeared at the U-Publish.com Web site at the end of 2001.

MightyWords and iPublish Close

Poynter and Snow have long maintained that while e-Books are almost certain to play a major role in the future of publishing, print-based technologies like POD and PQN will dominate the industry for several more years. Recent announcements about closures at two of the largest digital-only web sites support our view. See items below for details.

iPublish.com Folds

According to Steven Zeitchik, a reporter for Publishers Weekly (12/10/2001) "Trade publishing's most elaborate experiment in e-publishing came to an end last week when Time Warner Trade Publishing announced it is folding iPublish and at least temporarily abandoning its idea of using the Web as a place to troll for unknown writers.

"The company will continue reprinting e-Book editions of paper books, and possibly even original work by print authors, via BookMark, the house's online marketing division. A transition team will stay on for one to two months to work on that integration.

Of the nine iPublish authors whose books were scheduled for print publication, the company expects most to still be published, possibly as part of the Warner mass market division. The iPublish Web site will be closed."

We believe that iPublish failed because it focused almost exclusively on e-publishing, and ignored the preference of today's readers for printed books.

Co-Author Danny O Snow met with Gregory Voynow of iPublish prior to its launch, and proposed publishing POD paperback editions of at least one title per month, but Snow's proposal went unanswered. Now, only a handful of printed books will survive iPublish, after a loss of $13 million.

MightyWords Shuts Down

In a related story covered by Edward Nawotka of *Publishers Weekly* (12/17/2001) "MightyWords.com, which created quite a bit of fanfare when it launched in March 2000 with plans to digitally distribute original short works, is closing down. CEO Chris MacAskill told PW 'The motivation for closing wasn't that we're running out of cash. It's that the adoption for digital publishing isn't happening as fast as we hoped.' He said the company has about a year and a half of working capital left.

MacAskill added, "the only digital publishers that are doing pretty well are those that started in digital and then went into print."

Chapter 11: Afterword

WHAT'S AHEAD? In spite of false starts, unrealistic expectations and plain ol' bad luck during the early years of e-Books, the inherent power of e-publishing remains strong. The benefits of producing books with minimal production and shipping costs, no warehousing expense or inventory tax, and unlimited 24/7 availability still virtually insure that e-Books will be an increasingly important part of publishing in the future.

The challenge will be to give readers a wide selection of quality reading material that is easy and affordable to use.

Achieving this goal will require that publishers embrace new business models, in order to build a financial incentive to release more good books in electronic form. It's clear by now that millions of consumers won't pay high prices for e-Book reading devices, or for e-Books themselves.

Alternative marketing models are plentiful. On the hardware side, look at cell phones: manufacturers give the hardware away in order to sell air time, and it works. The same kind of strategy might work for e-Books. On the content side, we've already discussed both inexpensive pay-per-read strategies, and "all you can eat" subscription models. Others are out there, waiting to be tested, as we are doing here.

Publishers also need to overcome their fear that affordable e-Books will erode the sales of tree-Books. That isn't going to happen any time soon.

Experiments like this one, though admittedly radical, might even show that e-Books actually *catalyse* the sale of printed books. But even if no one buys the paperback edition of *Steal this e-Book!* in the months ahead, it won't mean that readers prefer e-Books to tree-Books. More likely it will mean that the audience for this narrow subject does.

Some of the eventual solutions may involve new hardware, software or copyright laws, but my prediction is that real answer will come from better marketing methods.

The changes required may sound scary to publishers at first. Early in 2002, the publishing industry might still be compared to a group of children contemplating a dive into unknown waters. Each child is saying "You go first!" In my view, it's only a matter of time before a few brave souls take the plunge. The others will follow when they see that the water's fine.

Cynics may argue that the failure of early e-Book libraries to sell meaningful numbers of books proves otherwise. But let's be frank; unlike Project Gutenberg, Online Originals, Peanut Press and a handful of others, many early commercial e-libraries were little more than online vanity presses, which would release almost anything—usually at the writer's expense. But no one likes bad books, and it's no surprise the public didn't buy many of them.

Real publishers, on the other hand, have vast stores of quality reading material that could be made available in electronic form. But the major players priced the hardware and the content too high in the early years, with predictably poor results.

Quality, variety, price and ease of use must all coalesce before the true potential of e-Books can be realized. It simply won't happen until then.

To paraphrase from Joseph Heller's *Catch-22*, there's a scene where the quintessential wartime entrepreneur, Milo Minderbinder, corners the market on Egyptian cotton—only to discover that no one will buy it. In desperation, he tries to get his friend Yossarian to eat a chocolate-covered cotton ball, to test the market for selling more of them to the Army. "But Milo, the men won't like this," says Yossarian, "Cotton is inedible."

"But the men *have to* like it," cries Milo. "You have to *make them* like it."

Sounds absurd, but today's readers are no more likely to buy millions of expensive, hard-to-use e-Books than to eat millions of cotton balls. The results are already in; it's pointless to try force-feeding the public something they clearly won't swallow.

New ways of making good, convenient electronic texts available at a low cost can be found. This will require some ingenuity, and some courage, from authors and publishers.

Meanwhile, I hope that experiments like this one, however modest or impertinent, will nudge the future of the "printed" word another baby-step in the right direction. You, Gentle Reader, can play your part too. Whether you want to read these words on a screen or on paper is up to you. But please, pass 'em along.

-- DOS, March 2002

Chapter 12: Resources

B ELOW, READERS WILL FIND a small sampling of Web sites that focus on e-Books, devices for reading them, and related subjects. Naturally only a fraction of those available in 2002 can be mentioned, and many more will appear in the years ahead:

ALEX Catalog of Electronic Texts
www.infomotions.com/alex
"Free ebooks in several formats."

Amazing Web Tales
www.webtales.com
"Read... Write... PUBLISH!"

Beehive Microtitles
http://microtitles.com

Being Digital
by Nicholas Negroponte
http://mitpress.mit.edu/bookstore/nonpress/being.html

BiblioBytes
www.bb.com
"A large selection of free electronic books."

BookTech
www.booktechmag.com
"The Magazine for publishers."

Catch Word, Ltd.
www.catchword.co.uk
"Electronic publishing solutions."

Dynamic Digital Content, Inc.
www.dynamicdc.com

eBookagent
www.eBookagent.net
e-Book Placement and Promotion."

eAuthorsOutlet
www.eauthorsoutlet.com
"The electronic self-publisher's marketplace."

e-Book Ad Daily News
www.ebookad.com

eBook Connections
www.ebookconnections.com

e-Book Newsletter
www.e-BookNewsletter.com
"News and resources for publishers, distributors, authors and
bookstores."

e-Books U.K
www.e-Books.uk.net
"Writers receive 80% of revenues."

eBookWeb
www.ebookWeb.org
"A leading resource for news about electronic books and devices for
reading them."

Electric Works Publishing
www.electricpublishing.com
"Publisher of e-Books since 1995."

Electron Press
www.electronpress.com
"We publish books on the Internet."

ePubZine
www.epubzine.com

"How-to, News, Reviews and Free Software for e-Book Publishers and Authors."

Independent Publisher Online
www.independentpublisher.com
"Leading the world of bookselling in new directions."

Infinity Publishing
www.InfinityPublishing.com
"Where the dream thrives…"

Internet Writer
www.internetwriter.co.uk
"The Internet: A Writer's Guide"

Internet Publishing Magazine
www.ipubmag.com

iUniverse
www.iuniverse.com
"The digital content leader."

Ivan Hoffman
Internet Law, Publishing Law, Copyrights, Trademarks, etc.
www.ivanhoffman.com

National Ass'n of Independent Publishers
www.publishersreport.com
"Dedicated to the idea that independent publishing is one of the last bastions of free speech."

Online Originals
www.onlineoriginals.com
"An online book publishing company specializing in new and original works in the arts and humanities."

Open eBook Forum
www.openebook.org

Para Publishing
Primary site of Dan Poynter
www.parapublishing.com
"The definitive location for book writing, independent publishing and promoting resources."

Peanut Press—Palm Digital Media
www.peanutpress.com

Planet e-Book
www.planetebook.com

Print Media Magazine
www.printmedia.com

Project Gutenberg
www.gutenberg.org
http://promo.net/pg

Publisher's Report
www.publishersreport.com
"Newsletter of the National Association of Independent Publishers"

Softskull Press
www.softskull.com
"Radically intelligent books."

Suite101
www.suite101.com
"The online publishing community of real people helping real people."

Texterity
Textcafe and e-Book Logistics
www.texterity.com

The Publishing Law Center
www.publaw.com

U-Publish.com
"How 'U' Can Compete with the Giants of Publishing"
by Dan Poynter and Danny O Snow
www.u-publish.com

UVA e-Book Library
http://etext.lib.virginia.edu/ebooks/ebooklist.html
"For Microsoft Reader and Palm Devices."

Wiley and Sons
www.wiley.com
"An independent, global publisher of print and electronic products, specializing in scientific and technical books and journals, professional and consumer books and subscription services, and textbooks and educational materials for colleges and universities."

About the Book

S*teal this e-Book!* is an irreverent collection of articles and letters about electronic publishing by Danny O Snow, with additional contributions by Richard Eoin Nash, Dan Poynter, Wade Roush and Glenn Sanders. It traced the evolution of e-Books in "ancient times," 1999 to 2002. Additional reports followed, as shown at:

www.u-publish.com/media.htm

In early reports, Snow flatly rejected the notion that challenges in DRM ("digital rights management," also called copy protection, encryption, etc.) should prevent authors and publishers from earning profits from e-Books. Instead, he argued that booksellers would need to update their technology and business models to meet the normal buying and reading patterns of consumers. He also reaffirmed his long-standing conviction that the inherent power of electronic publishing virtually insured e-Books an important role in the future of publishing.

About the Writer

Harvard graduate Danny O Snow, also co-author of the book titled *U-Publish.com* with Dan Poynter, has been widely quoted about new publishing technologies by many broadcast media such as AP, NPR, UPI, the nationally syndicated "Ask Heloise" radio program, and Talk America Radio. He has also been quoted by scores of print media including *The Los Angeles Times, BookTech Magazine, Publishers Weekly, The Wall Street Journal, Washington Times* and many others. Snow is known as one of the first print publishers to embrace e-Books, blogging, podcasting and "Print on Demand" book publishing. He has also served as a panelist and moderator at national publishing events such as the North American Publishing Company's "PrintMedia" expos and PMA's "Publishing University," as a POD book publisher with Unlimited Publishing LLC, and as a contributing editor for *BookTech the Magazine* for publishers. In 2008 he was named a senior fellow of the Society for New Communications Research (Palo Alto, California, USA), a global think tank dedicated to the advanced study of new and emerging media.

www.ingramcontent.com/pod-product-compliance
Lightning Source LLC
Chambersburg PA
CBHW022128170526
45157CB00004B/1788